Regenerative Engineering Society 2019

Held at the 2019 AIChE Annual Meeting

Orlando, Florida, USA
10-15 November 2019

ISBN: 978-1-7138-0556-4

Regenerative Engineering Society 2019

Held at the 2019 AIChE Annual Meeting

Orlando, Florida, USA
10-15 November 2019

AIChE
The Global Home of Chemical Engineers

Printed from e-media with permission by:

Curran Associates, Inc.
57 Morehouse Lane
Red Hook, NY 12571

Printed with permission by Curran Associates, Inc. (2020)

For permission requests, please contact AIChE
at the address below.

AIChE
120 Wall Street, FL 23
New York, NY 10005-4020

Phone: (800) 242-4363
Fax: (203) 775-5177

www.aiche.org

Additional copies of this publication are available from:

Curran Associates, Inc.
57 Morehouse Lane
Red Hook, NY 12571 USA
Phone: 845-758-0400
Fax: 845-758-2633
Email: curran@proceedings.com
Web: www.proceedings.com

TABLE OF CONTENTS

581719 Quantum-on-Demand Biofabrication of 3D Tissue Test Systems

Scientific modeling can be used to conceptualize, test, understand, and direct complex natural systems. The ability to create accurate models of three-dimensional (3D), functioning tissues is a growing reality. Benchtop tissue models, i.e. tissue test systems, have many potential advantages over current modeling technologies, including two-dimensional (2D) cell culture and animal models, for biomedical research. Three-dimensional fabricated tissues are more representative of natural tissues than 2D cell cultures, are easier to control, replicate, and discern than animal models, and are potentially a cost saving for biomedical research. Quantum-on-demand biofabrication, i.e. the deposition of biomaterials such as cells and granules in a specific quantity at a specific location, offers control over cell distribution, biomolecule deposition, and biomaterial geometries and surfaces. Tissue models can be used to investigate tissue function and disease progression, discover new pharmaceuticals, teach, and personalize treatments for patients.

580862 Robustness of Cellular Signaling with Respect to Dosage

Introduction: How tissues are patterned in a natural context can serve as the foundation for our future methods of tissue engineering and regeneration. In a developing tissue, the timing and placement of gene expression is crucial to the proper assignment of cell fate, and subsequent differentiation of the cells into a reproducible pattern of tissues and organs. Proper gene expression patterns are often achieved through the action of spatial gradients of chemicals, typically a class of proteins called morphogens, which signal to cells in a concentration-dependent fashion. However, morphogen-mediated patterning models face a paradox: spatial patterns of gene expression are surprisingly robust to variations in morphogen dosage despite the fact that, by definition, cells must be sensitive to morphogen concentration. In this work we investigate the robustness of the Dorsal/NF-κB signaling module with respect to perturbations to the dosage of maternally-supplied *dorsal* (*dl*). Through the action of Toll receptor signaling, which is high on the ventral side of the embryo, Dorsal (Dl) becomes free of its cytoplasmic inhibitor Cactus (Cact)/IκB and is able to enter the nucleus to regulate gene expression[1]. In this manner, a spatial gradient (ventral high, dorsal low) in Dl nuclear concentration is established, which acts to pattern the dorsal/ventral (DV) axis of the early embryo. Although these biophysical interactions have been characterized in detail, how their function is integrated to produce robust gene expression patterns to variations in the *dl* maternal dosage is still unclear.

Results: We measured gene expression of two Dl target genes: *sna*, which is on the ventral-most 20% of the embryo, and *sog*, which resides in a broad lateral region, in embryos carrying 1, 2, or 4 copies of *dl*. The gene expression boundaries have small, but detectable differences depending on *dl* dosage. However, naïve models of Dl distribution, which assume that Dl concentration globally scales with *dl* gene dosage, predict high sensitivity gene expression boundaries. This implies there must be mechanisms that alter the shape of the Dl gradient when *dl* dosage is perturbed. Previous work[2,3] has suggested three such mechanisms: facilitated diffusion of Dl by Cact, saturation of active Toll receptors, and the presence of Dl/Cact complex in the nucleus. We perform a parameter screen using a mechanistic model of the Dl/Cact system to show that these mechanisms are together necessary for robustness of gene expression boundary with respect to *dl* dosage.

Conclusions: In the Dl gradient system, gene expression is robust due to the action of three mechanisms: facilitated diffusion, Toll receptor saturation, and the action of Dl/Cact complex in the nucleus. Morphogen gradients must be able to determine cell fates in a concentration-dependent fashion, while at the same time maintain robustness with respect to morphogen dose. The Dl/Cact system shows an example of this engineering principle. Ultimately, we may be able to use our engineering knowledge gained from the basic study of tissue patterning to benefit the field of regenerative medicine.

This research was supported by NSF CBET-1254344 and NIH R21HD092830.

References:

1. Reeves, GT & Stathopoulos, A. *Cold Spring Harb Perspect Biol* (2009) **1**: a000836.
2. Carrell, SN et al. *Development* (2017) **144**: 4450-4461.
3. O'Connell, MD & Reeves GT. *PLoS Comput. Biol.* (2015) **11**: e1004159.

564816 Synergistically Enhance Bone Cell Proliferation and Osteogenesis on 3D-Printed Scaffolds Using Two-Dimensional Black Phosphorus and Graphene Oxide Nanosheets

Synergistically Enhance Bone Cell Proliferation and Osteogenesis on 3D-Printed Scaffolds Using Two-Dimensional Black Phosphorus and Graphene Oxide Nanosheets

Xifeng Liu[ab], A. Lee Miller II[b], Sungjo Park[c], Matthew N. George[a], Brian E. Waletzki[b], Haocheng Xu[a], Andre Terzic[c], and Lichun Lu*[ab]

a Department of Physiology and Biomedical Engineering, Mayo Clinic, Rochester, MN 55905, USA

b Department of Orthopedic Surgery, Mayo Clinic, Rochester, MN 55905, USA.

c Department of Cardiovascular Diseases and Center for Regenerative Medicine, Mayo Clinic, Rochester, Minnesota 55905, USA

* Corresponding author. Email address: Lu.Lichun@mayo.edu; Phone: 507-284-2267; Fax: 507-284-5075.

Keywords: 2D Materials; 3D-Printed Scaffolds; Bone Cell Osteogenesis

1. Introduction: Incorporation of two©\dimensional (2D) materials to tissue scaffolds for improved biocompatibility and cell affinity has emerged as a promising new strategy in tissue engineering. For example, 2D graphene oxide (GO) has been widely reported to enhance cell adhesion and proliferation. Recently, a newly emerged black phosphorus (BP) 2D material has attracted attention in biomedical applications due to its unique mechanical and electrochemical characteristics. In this study, we investigated the synergistic effect of these two types of 2D materials on cell osteogenesis on three-dimensional (3D) printed poly(propylene fumarate) (PPF) scaffolds for bone tissue engineering (Fig. 1a).

2. Materials and Methods: Graphene oxide sheets were oxidized and exfoliated from natural graphite (~150 mm flakes) through improved Hummers¡¯ method.[1] BP nanosheets were synthesized using a liquid exfoliation method. Poly(propylene fumarate) was synthesized and used as the cross-linkable polymer for 3D printed scaffolds.[2] The 3D scaffold model was designed using computer-aided design (CAD) software (SolidWorks Corp., Concord, MA) and printed using the Viper si2 stereolithography system (3D Systems, Valencia, CA).[3] The ammonolyzed 3D-PPF-Amine scaffolds were coated with 2D materials by soaking in solution containing BP nanosheets, GO nanosheets or both BP and GO nanosheets. Sterilized functionalized 3D scaffolds were adhered to the bottom of 48-well tissue culture polystyrene (TCPS) plates and MC3T3 pre-osteoblast cells were seeded. Upon harvest, cells were fixed in paraformaldehyde (PFA, 4% solution) and stained with anti-vinculin-FITC antibody (Sigma-Aldrich Co., Milwaukee, WI) to label vinculin and rhodamine-phalloidin (RP, Cytoskeleton Inc, Denver, CO, USA) to stain cellular filaments. The immune-fluorescence stained cells on the functionalized 3D Scaffolds were immediately scoped on an inverted laser scanning confocal microscope (Carl Zeiss, Germany).

3. Results and Discussion: All 3D scaffolds after coating with both types of 2D materials showed good biocompatibilities (Fig. 1b). In addition, the increased surface area provided by GO sheets enhanced cell attachment at the initial stage. Afterward, cells were further stimulated by phosphate, which was continuously released from slow oxidation of BP nanosheets and played an important role in osteoblast differentiation. Through the use of 3D confocal imaging, unique interactions between cells and BP nanosheets were observed, including a stretched cell shape and development of filaments around the BP nanosheets, along with increased cell proliferation when compared with scaffolds incorporating only one of the 2D materials (Fig. 1c). All these results indicate that the incorporation of 2D BP and GO materials could effectively and synergistically stimulate cell proliferation and osteogenesis on 3D tissue scaffolds.

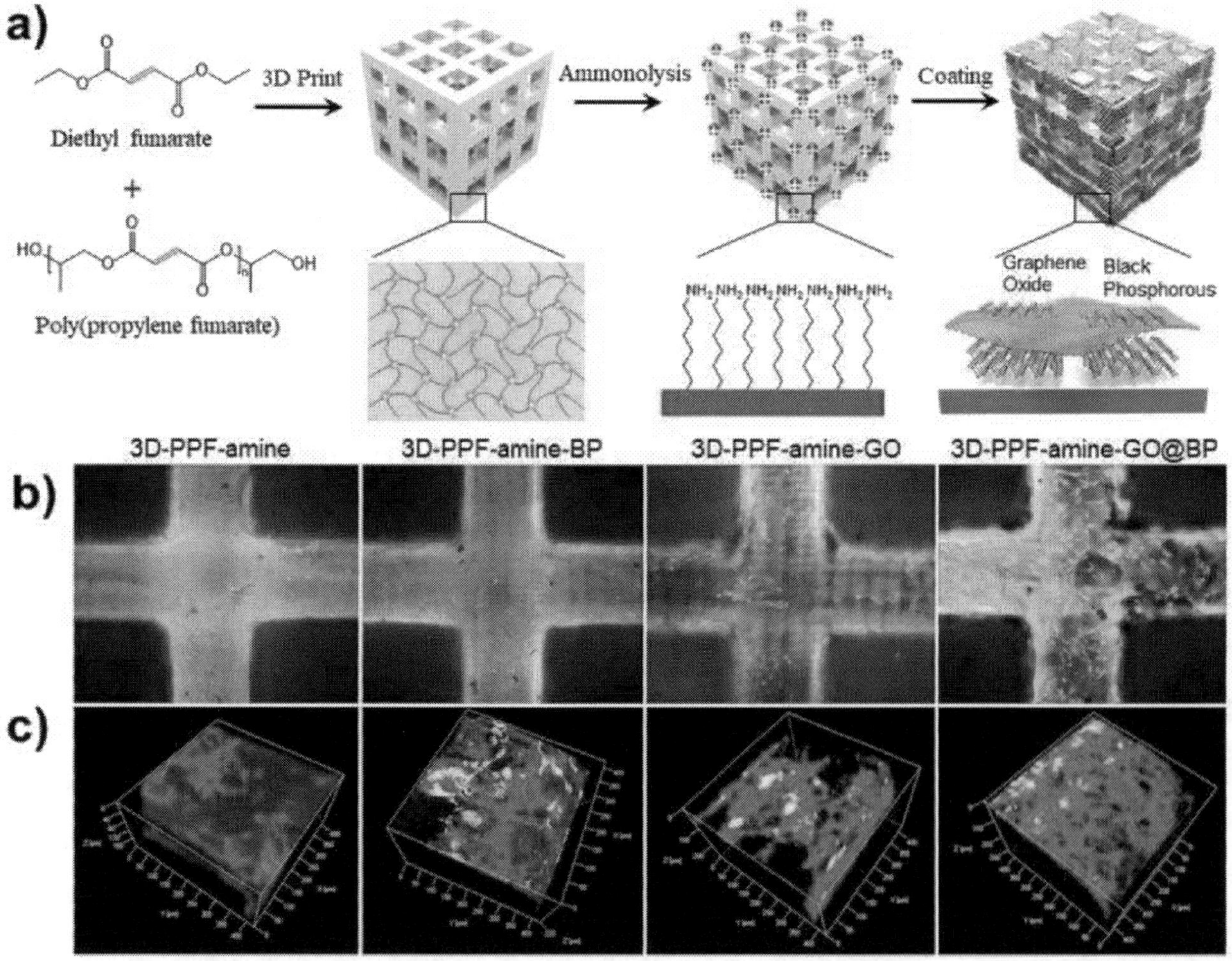

Fig. 1 a) Fabrication process of 3D printed scaffolds functionalized with 2D GO and BP nanosheets. b) Live/dead staining of ammonolyzed 3D scaffolds as well as scaffolds functionalized with BP, GO and GO@BP nanosheets. c) Immunofluorescence images of MC3T3 pre-osteoblasts growing on functionalized 3D scaffolds.

4. Conclusions: We fabricated a series of 3D-printed scaffolds functionalized with 2D materials for use in tissue engineering. Our results showed that the addition of 2D BP and GO nanosheets to scaffolds improved cell adhesion, proliferation, and differentiation.

Acknowledgments: This work was supported by National Institutes of Health grant R01 AR56212.

References

[1]. X. Liu *et al*, Journal of Materials Chemistry B 2016, 4 (43), 6930-6941.

[2]. X. Liu *et al*, RSC Advances 2015, 5 (27), 21301-21309.

[3] K. W. Lee *et al*, Biomacromolecules 2007, 8 (4), 1077-1084.

567121 Biological Response Precision-Controlled By Spherical Pores

Implanted biomaterials, after approximately one month, are observed in an avascular, collagenous capsule – this outcome is referred to as the foreign body reaction (FBR). This reaction starts with recruited macrophages that then fuse into giant cells leading to the avascular, fibrotic healing. Such healing can be undesirable for implanted sensors, electrodes, drug delivery systems and other devices. We have developed a number of strategies based on polymeric materials to address this FBR. One such strategy will form the focus of this talk. We have fabricated hydrogel materials with uniform, interconnected 30-40 micron pores by a sphere-template method -- upon implantation the pores are infused by macrophages. The macrophages within the pores orchestrate a different healing process from that seen with non-porous materials or materials with non-optimal pore sizes. Macrophages are biomechanically directed to a unique phenotype that encourages reconstructive healing (M2). Thus, the inflammatory process is exploited to generate improved healing – a new biocompatibility. Reconstructive healing has been noted subcutaneously, percutaneously and within heart muscle, cornea and bone. The surprising and desirable outcomes from these implanted porous structures led us to commercialize them. The path from laboratory curiosity to a product and medical device that is now used in humans will be discussed.

569596 Correlation between Fluid Shear Stress and Morphological Behaviour of Valvular Endothelial Cells

Valvular endothelial cells (VECs) line the outer layer of heart valves and withstand oscillatory and laminar shear forces along with other mechanical stimuli. Compared to vascular endothelial cells, VECs are understudied and as a consequence their exact physiological behaviour is unclear. Thus, endothelial dysfunction mechanisms leading to valve sclerosis or calcification are yet unknown. In vitro studies of VECs cultured under physiological mechanical stimulation can uncover mechanisms of VEC mechanobiology. Previous studies have shown that vascular endothelial cells align parallel to fluid flow while, in contrast, VECs align perpendicular to blood flow. Recent work from our group has demonstrated that this behaviour varies for different ranges of shear stress values. Here, we aim to show that VEC behaviour does not depend exclusively on fluid shear. There may be several other factors influencing such behaviours, such as confluence levels, time exposed to shear, as well as the stiffness of the supporting scaffold.

This work is based on the study of behaviour of VECs under different shear stress levels and different times of exposure. VECs are cultured on soft substrates on a custom-made parallel-plate chamber. This chamber is specially designed for cultures on soft scaffolds while applying fluid shear on cells in monolayer. Polydimethylsiloxane gels are prepared in the chamber and have their surfaces modified using sulfo-SANPAH as a linker for collagen type I. After collagen treatment, 10^4 VECs/cm^2 are seeded on the surface and allowed to adhere for 48 h without shear exposure. Upon cell adhesion, 2.6, 6.5, 13 and 20 dyne/cm^2 fluid shear are applied on the monolayer culture for either 24 h or 48 h. Cultures then undergo immunofluorescence processing prior to imaging and image analyses of cell aspect ratios, circularity, area, alignment, etc.

Our results demonstrate that with increasing fluid shear, VECs tend to orient in more diverse way, losing their traditional cuboidal morphology, as quantified by cell aspect ratio and circularity parameters. In addition, higher shear levels decrease VEC area, regardless of confluence level. In addition, we observed that, the threshold for cell alignment with shear is much higher than previously observed. High shear also increases VEC differentiation and mesenchymal transition, as validated by F-actin and interstitial cell phenotypes. These results increase the current understanding of VEC physiology and this knowledge is key for the future engineering of replacement heart valves.

578181 Metabolite-Based Modulation of Dendritic Cells for Developing Effective Immunotherapy

INTRODUCTION – Aging and diabetes associated chronic / slow healing cutaneous wounds have hallmark immune signatures, including *delayed* T-cell and macrophage infiltration, and increased pro-inflammatory immune cell infiltration for *extended periods* (compared to non-aged control). *If there were a method to modulate the local immune responses in a temporal manner then it would be possible to accelerate the wound healing of cutaneous wounds in aged and/or diabetic immune system.* Notably, metabolic pathways control immune cell functions, and modulating these pathways can determine overall immune system response. Therefore, in this work, polymers were generated with metabolite monomers that can deliver metabolites and directly modulate the function of immune cells.

RESULTS – This work for the first time (to the best of our knowledge) demonstrates utilization of metabolites to modulate immune-metabolic responses in wound healing. NMR and FTIR studies were performed to characterize the polymers. Moreover, oil in water emulsions were utilized to generate microparticles of these metabolite-based polymers. These microparticles are capable of encapsulating drugs and release them in a sustained manner (shown via representative drug-molecule rhodamine). The release kinetics of metabolites from these polymers were determined using NMR. Moreover, degradation kinetics of the polymers was determined using weight loss experiments. Importantly, it was demonstrated that these polymers could modulate the function of mouse bone marrow derived dendritic cells and allogenic T-cells in a mixed lymphocyte reaction. Lastly, we demonstrate that these polymers when applied in the presence of tegaderm were able to modulate the wound healing rate in BALB/c mice.

CONCLUSION - In summary, metabolite-based polymers were generated that were capable of controlling the pro- or anti-inflammatory responses of immune cells *in vitro* and *in vivo*. Notably, these polymers were able to lead to healing of cutaneous wounds in mice.

562790 The Effect of Centrifugal Force on the Mechanism of Wharton's Jelly Mesenchymal Stem Cells Transfection

A human mesenchymal stem cell (hMSC) is an adult stem cell characterized by: being an undifferentiated cell, having adapted to a specific growth environment, producing a large number of differentiated and functional progenies, and regenerating tissues after injury or lesion. All those characteristics make them a promising cell source for regenerative medicine, especially for their multipotent capacity which render them capable of differentiating into multiple lineages including cartilage, adipose, and bone. Although controlling the phenotype of MSCs is a central challenge in tissue engineering, gene therapy has become more attractive since with the introduction of exogenous genes into the cell, the latter is expected to produce its own necessary growth factors and cytokines for its differentiation. This approach would be preferable over the delivery of recombinant cytokines and growth factors, as it involves the administration of non-physiological concentrations with a short half-life, fast body clearance and a lower therapeutic effect in comparison to natural proteins.

In gene therapy, viral vectors such as retrovirus, lentivirus and adenovirus, have been used for the delivery of genes into cells via a process known as transduction. Even though it offers high gene delivery (nearly 99%) and stable gene expression, many limitations remain associated with viral vectors such as insertional mutagenesis, immunogenicity, limited DNA packaging capacity and cumbersome large-scale production. Non-viral gene carriers are promising alternatives for gene delivery and have the potential to address these limitations. Moreover, the transient expression associated with these systems can be more compatible with the natural wound healing processes. However, transfection has significantly lower gene entry rates compared to the viral transduction (approximately 10% to 30% in hMSC). This lower efficiency is due to the mechanism of transfection, which has been theorized to be affected by multiple factors regarding nanoparticle properties, including: size, charge, degradation, endosomal escape and arrival to the nucleus. These barriers depend directly on the chemical nature of the material that covers the gene and its non-specific interaction with all the pathways within the cell. Therefore, there is still need for more in-depth studies, with the aim of improving transfection techniques using this type of vectors or physical mechanisms.

One of the most promising non-viral gene carriers is polyethylenimine (PEI). This polymer, which has a great variety of molecular weights and degrees of branching, has been extensively studied *in vitro* ever since its initial use as gene carrier. PEI has an advantage over other polycationic polymers in that it acquires a remarkable DNA condensation capability and a potential ability to protect DNA from enzymatic degradation. In addition, the three different types of amines that it has are responsible for generating the buffering capacity over a wide range of pH values. This facilitates endosomal escape of carriers, which has been named the "proton-sponge" effect.

As discussed before, transfection presents a poor efficiency of gene uptake. However, we tested here the hypothesis that the efficiency of gene transfection is limited by a simple physical barrier: low concentration of gene (DNA) on the surface of the cell. Therefore, the present study evaluated the mechanism of centrifugation during the transfection process in order to increase the concentration of DNA on the cell surface, thus generating higher transfection efficiencies. The present study evaluated the mechanism of centrifugation during the transfection process in order to increase the concentration of DNA on the cell surface, which allowed higher transfection efficiencies. For this purpose, mesenchymal stem cells derived from umbilical cord, reporter genes such as GFP and luciferase and, branched PEI 25kDa were used to evaluate and optimize the transfection technique by assessing various centrifugal forces. We were able to standardize a transfection protocol in which the weight ratio and different concentrations of complex were evaluated to obtain transfection rates above 20% with a non-viral method.

569466 Aggregation of Mesenchymal Stem Cells Rejuvenates Functionality Following *in Vitro* Aging through Induction of the Integrated Stress Response

Mesenchymal stem cells (MSCs) have shown promising therapeutic potential due to their impressive ability to secrete inflammatory modulatory, angiogenic, and regenerative cytokines along with their ability to differentiate and aid in regeneration. Most work with MSCs though has been done in model systems requiring only a small amount of MSCs. Scaled-up expansion of isolated MSCs is carried out on stiff culture plastic which increases the surface modulus one thousand-fold [1]and limits cell-cell interactions. This type of culture results in *in vitro* aging characterized by increased senescence, decreased glycolytic metabolism, reduced autophagic activity, and loss of multipotent differentiation potential [2]. For these reasons understanding the mechanism of culture induced changes is critical for maintaining MSC phenotype during expansion and proper implementation of MSC therapy.

Protein homeostasis is critical for cells functional maintenance and retention of stemness. As cells age, they loss control over proteasome activity leading to the accumulation of misfolded proteins that would normally be handled by the integrated stress response (ISR). Normally, when misfolded proteins occur they are coated in chaperone proteins such as binding immunoglobin protein (BiP). BiP normally prevents activation of eukaryotic initiation factor 2 alpha kinase 3 (PERK), but as soluble BiP decreases PERK dimerizes/auto-phosphorylates and signals a reduction in protein initiation complex activity causing a cessation of protein synthesis. The misfolded protein is then ubiquinated and lysed. Proteasome autophagy is especially critical in maintaining cell stemness and fighting senescence. It has recently been shown that when autophagy was inhibited in naïve muscle-stem cells they took on the phenotype of *in vitro* aged cells, and when aged cells had autophagy restored they re-established their naïve phenotype [3]. Previously, it has also been shown that embryonic stem cells hyper-control their proteome to maintain stemness by limiting initiation, increasing folding through chaperone protein content, and increasing proteasome activity to degrade build-up of misfolded proteins and prevent defects from going into daughter cells.

It has previously been shown that aggregation of MSCs leads to heightened autophagy [3], and a reversal of metabolic phenotype back to aerobic glycolysis [4]. For this reason, we have explored what effects aggregation has on the proteome to determine what pathways are elevated by aggregation in bone-marrow derived MSCs (bMSCs), adipose derived MSCs (ASCs), and dermal fibroblasts (FB). Pathway analysis showed a decrease in the eIF2 pathway; however, western blot and ingenuity pathway analysis revealed that bMSC and ASCs had augmented eIF2α and eIF2γ. Specifically, eIF2α was significantly phosphorylated and eIF2γ was decreased showing heightened initiation control following aggregation. Since phosphorylation of eIF2α leads to the subsequent translation of activating transcription factor 4 (ATF) which in turn starts a cascade increasing autophagy. Measured levels of ATF in bMSCs and ASCs decreased following aggregation, while FB didn't shift. Downstream proteins DNA-damage inducible transcript 3 (CHOP) showed no shift though autophagy substrate microtubule associated light chain protein 3B (LC3B) showed a significant upregulation following aggregation in bMSCs and ASC. CHOP is also known to regulate growth arrest and DNA-damage inducible protein (GADD34), which is known to dephosphorylate eIF2α, showed lower levels in bMSC and ASC aggregates. This would lock the aggregated stem cells into a higher basal ISR state and recovering the *in vivo* homeostatic state. Further proteomic analysis revealed that MSCs and ASCs showed heightened proteasome control, while FB had higher chaperon content showing a distinct difference in stem and differentiated cells ISR mechanism.

This work has illustrated the importance of 3D culture in the form of aggregation for the maintenance of MSC stem cell properties. This was carried out in MSCs by induction of the ISR pathway. The ISR pathway *in vivo* allows cells to mitigate cellular stress, in particular stress associated with protein misfolding, which is a problem commonly encountered during *in vitro* aging. Aggregation of FB did not result in the same trend of ISR upregulation suggesting a loss of sensitivity following differentiation. Aggregation induce ISR in MSCs lead to heightened basal levels as the *in vivo* tempering protein GADD34 was not upregulated, this heighten basal ISR also lead to increased autophagy and increased stem protein levels. Taken together these results show a promising alternative to traditional expansion culture for MSCs which leads to loss of function. This will allow for more effective scale-up thereby increasing the therapeutic efficacy of MSC treatments.

References

[1] Z. Cesarz, J. L. Funnell, J. Guan and K. Tamama, "Soft Elasticity-Associated Signaling and Bone Morphogenic Protein 2 Are Key Regulators of Mesenchymal Stem Cell Spheroidal Aggregates," *Stem Cells and Development,* vol. 25, no. 8, pp. 622-35, 2016.

[2] A. Banfi, A. Muraglia, B. Dozin, M. Mastrogiacomo, R. Cancedda and R. Quarto, "Proliferation kinetics and differentiation potential of ex vivo expanded human bone marrow stromal cells: Implications for their use in cell therapy," *Experimental Hematology,* vol. 28, no. 6, pp. 707-15, 2000.

[3] L. Garcia-Prat, M. Martinez-Vicente, E. Perdiguero, L. Ortet, J. Rodriguez-Ubreva, E. Rebollo, V. Ruiz-Bonilla, S. Gutarra, E. Ballestar, A. L. Serrano, M. Sandri and P. Munoz-Canoves, "Autophagy maintains stemness by preventing senescence," *Nature,* vol. 529, no. 7584, 2016.

[4] R. Pennock, E. Bray, P. Pryor, S. James, P. McKeegan, R. Sturmey and P. Genever, "Human cell dedifferentiation in mesenchymal condensates through controlled autophagy," *Scientific Reports,* 2015.

[5] Y. Liu, N. Munoz, A.-C. Tsai, T. M. Logan and T. Ma, "Metabolic Reconfiguration Supports Reacquisition of Primitive Phenotype in Human Mesenchymal Stem Cell Aggregates," *Stem Cells,* vol. 35, no. 2, pp. 398-410, 2017.

579923 Three-Dimensional Culture of Human Adipose-Derived Stem Cells By Surface Modification Using Elastin-Based Copolymers

During their differentiation and maturation into adipocytes *in vitro*, human adipose derived stem cells (hASCs) begin to accumulate lipids during maturation and begin to become more buoyant over time. Such fat-laden cells eventually lift off of the culture surface and are lost during the next media change. Because of the loss of physiologically relevant spheroids, elastin-like polypeptide-polyelectrolyte coated surfaces were used to form spheroids for long term culture, in order to provide a proven, long term culture method for mature adipocytes *in vitro*. While there are several methods of forming spheroids, there is not a general consensus on which methods are the most useful for stem cell proliferation, differentiation, and maturation ability. This is also true for most cell types, including adipocytes. Overall, our research provides a proven cell culture model for future studies involving long-term culture of mature adipocytes derived from hASCs. In addition to using elastin-like polypeptide-polyelectrolyte coated surfaces, alternative methods such as hanging drop, ultra-low attachment plate culture, and suspension cultures were compared for their ability to form and retain productive hASC spheroids. It was found that there are differences in spheroid formation and maturation within different molecular weights of the same polyelectrolyte, polyethyleneimine, as well as different compositions of the same molecular weight. Though the selected alternative spheroid forming methods were comparable to the elastin-like polypeptide-polyelectrolyte coated surfaces in terms of adipogenic maturation, the coated surfaces had greater amounts of spheroid retention, which allows for more physiologically relevant spheroids to be acquired.

575197 Delivering Therapeutic Drugs to Rejuvenate Progenitor Cells

The controlled delivery of bioactive molecules to the cell surface can significantly widen the therapeutic window, as well as enhance therapeutic potential of progenitor cells. Here we describe a strategy to rejuvenate endothelial progenitor cells (EPCs) via the conjugation of adjuvant drug-loaded nanoparticles to the surface of progenitor cells. Cord blood EPCs isolated from patients with gestational diabetes exhibit lower vasculogenic potential and altered epigenetic regulation. By providing sustained pseudoautocrine stimulation, we can significantly rejuvenate the progenitor phenotypes and improve therapeutic potential of EPCs. Collectively, strategy to deliver therapeutic drugs to the cell surface represents a powerful tool that has a wide range of translational applications.

573531 Peptoid-Based Substrates for Human Neural Stem Cell Differentiation

Pluripotent human neural stem cells can repair damage in the brain due to stroke or neurodegeneration. Neural stem cells will differentiate into different neural cell types depending on substrate and cell-to-cell interactions. We propose that peptoid-based substrates could be utilized to promote neural stem cell growth and differentiation. Peptoids are biomimetic poly-N-substituted glycines that are resistant to degradation by proteases. Peptoids are similar in structure to peptides, with the side chains bound to the backbone nitrogens rather than the alpha carbons. These structures have potential uses in biomedical applications due to peptoid stability under physiological conditions, customizable sequence design, and structural similarity to naturally occurring peptides. The tunability of peptoids allows for the addition of a myriad of side chains that could be added to achieve the desired function. In this study, a peptoid that self assembles into 2-4 micron diameter microspheres and a peptoid capable of engaging in non-metal catalyzed click chemistry was synthesized. Human neural stem cells were plated onto the peptoid-microsphere substrate and compared to commercially available substrates to differentiate neural stem cells. The capability of the click chemistry peptoid to attach growth factors in physiological conditions was also tested.

AUTHOR INDEX

AIChE
120 Wall Street, FL 23
New York, NY 10005-4020

CURRAN ASSOCIATES INC.
proceedings.com

ISBN 978-1-7138-0556-4
9 781713 805564